Communication of drones telemetry using a flight controller

Roland Büchi

Bibliografische Information der Deutschen Nationalbibliothek
Die Deutsche Nationalbibliothek verzeichnet diese Publikation in der Deutschen Nationalbibliografie; detaillierte bibliografische Daten sind im Internet über www.dnb.de abrufbar.

Impressum

Herstellung und Verlag: BoD – Books on Demand, Norderstedt
ISBN: 978-3-7562-0444-1

FSC
www.fsc.org
MIX
Papier aus verantwortungsvollen Quellen
Paper from responsible sources
FSC® C105338

Communication of drones telemetry using a flight controller

Communication of drones telemetry using a flight controller

The multicopters are particularly suitable for dealing with the topics of telemetry or photo flights with UAV, Unmanned Air Vehicles. They allow to float in one place. This also enables a particularly good data transfer to the ground. In addition, very good pictures can be taken and transmitted in this way. This publication deals with the transmission of telemetry and image data with an octocopter and their visualization at a ground station, using a flight controller.

1. Octocopter

For this purpose, a multicopter with 8 motors and propellers is built and used as a practical example, i.e. an octocopter. 8” propellers are used in combination with Hacker A20-50S motors. On the one hand, this enables a certain redundancy. Thus, even in the event of a motor failure, a reasonably safe landing is still possible. On the other hand, you can also achieve a good payload with small 8" propellers, which is about 500g in the example. The octocopter is not designed with the otherwise usual round arrangement of the propellers. Instead of that, two rows, each with four propellers, are mounted parallel to each other.

This arrangement is also called an H arrangement because it is similar to this letter. Depending on the battery and additional devices such as telemetry transmitter, camera and video link, the resulting octocopter has a take-off weight of around 1000g. The motors each produce about 3N or 300g of thrust. So that's a total of about 24N or 2400g. 3S LiPo's are used as batteries, with between 2600mAh and 3800mAh. The flight time is between 15 and 20 minutes.

Figure 1 shows the octocopter. It is equipped with telemetry. You can see, among other things: the antennas of the telemetry and video transmission, a small camera, mounted without a gimbal, a GPS receiver and the flight controller. These components and their evaluation at the ground station are described in more detail in the next chapters.

Figure 1: Octocopter in H- configuration.

2. Flight controller

Flight controller are the central control unit of these aircraft. Their main task is to control the motors and propellers in such a way that the desired flight phases can be carried out correctly. In the simplest case, this is a normal remote-controlled flight with axis stabilization. Depending on the setting, it can also be altitude control via an air pressure sensor, GPS-supported hovering at one point, a controlled drift function to the left or right or forwards or backwards, or even an independent flight based on previously programmed waypoints.

Most modern flight controller also support telemetry and First Person View, FPV. In addition, the flight controller also communicate directly with the receiver of the remote control. Figure 2 shows a schematic overview of such a system. Figure 3 shows a flight controller with its connections.

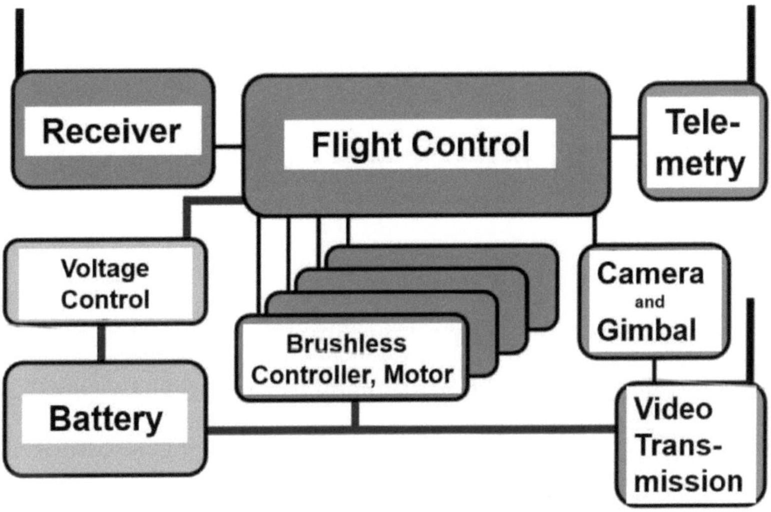

Figure 2: Schematic overview of a system with flight controller.

Figure 3: A flight controller with its connections.

This shows that the connections for all important functions such as telemetry or GPS are already available. These are also supported by the configurable software. In most cases, various options are supported for connecting the receiver. In this specific example, either the DSM signal supported by the Spektrum remote controls or the pulse-pause modulation signal (PPM) is required.

If the receiver cannot provide this signal, a PPM encoder can also be used. At its input are the PWM signals from up to eight channels. From this, an encoder composes the PPM signal, which is available at its output for the flight controller.

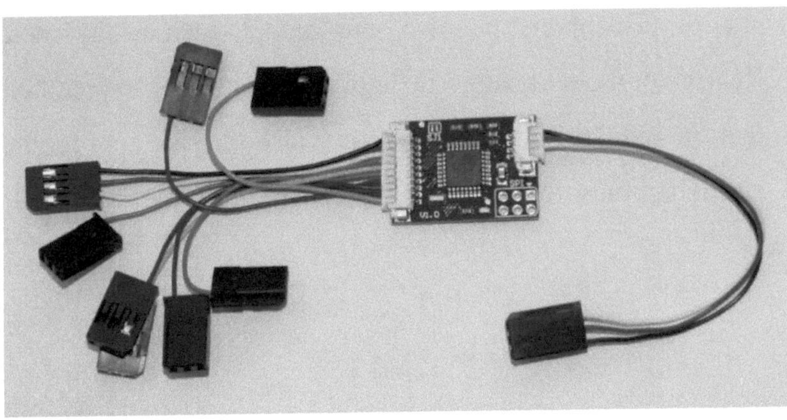

Figure 4: PPM- Encoder.

The flight controller outputs are visible in figure 3 at the front. In addition to eight PWM outputs for connecting brushless controllers to control the motors and propellers, additional servo connections are also provided. This can also be used to connect a motorized camera gimbal, for example. The necessary sensors for position detection of the multicopter are built into the flight controllers in order to keep it stable in the air and to manoeuvre. Therefore, the servos of a gimbal can also be controlled in such a way that the camera can always be held at the desired angle, regardless of which flight maneuvers the multicopter is currently performing. Such flight controllers are available today from various suppliers. With the appropriate configuration, it is often not only possible to control quadcopters, but also other models such as airplanes, RC cars or model boats. All flight controller have in common that, in addition to the sensors, they also provide appropriate configuration software with a USB interface to the PC.

3. Camera and video link

The octocopter discussed here has a simple built-in camera and video link. For weight reasons, the camera is mounted here without a gimbal. It is shown in figure 7. It is tilted slightly forward and rigidly mounted so that it can take good pictures with landscape and horizon. The resolution is 600 TVL, i.e. 600 horizontal pixels. It is therefore not suitable for taking high-resolution photos. However, it is a good choice for fast video transmissions to video goggles or a ground station. And it's also very light.

The video link is carried out with a 5.8 GHz transmission. Figure 5 shows this. Power is supplied via the balancer connection of the LiPo battery. This is freely available during the flight. On the ground, this image data can be made visible in various ways, for example with video glasses, which is described below. In this example, however, another possibility is shown first. The video glasses serve here only as a receiver of the image data. The video output is read into the PC with a so-called frame grabber.

A frame grabber converts the image data received via the receiver into data that can be transmitted via the USB interface. The Mission Planner is the software belonging to the flight controller used in the example. This allows various configurations or visualizations to be made. This is shown in figure 10 below. To avoid interference to the remote control, the video transmission is carried out at 5.8 GHz. The remote control transmits on 2.4 GHz.

Figure 5: Video transmission with 5.8 GHz.

The ground station thus consists of the PC, into which both the telemetry receiver and the frame grabber are plugged in via USB interfaces. The 'Mission Planner' software can display all data, GPS, images, battery voltage and more on one screen. This is also possible with other softwares for specific flight controller.

This also makes it possible to compile a complete overview of different data on the same screen. Thus, the status of the system can be monitored as well as possible during the flight, almost in real time.

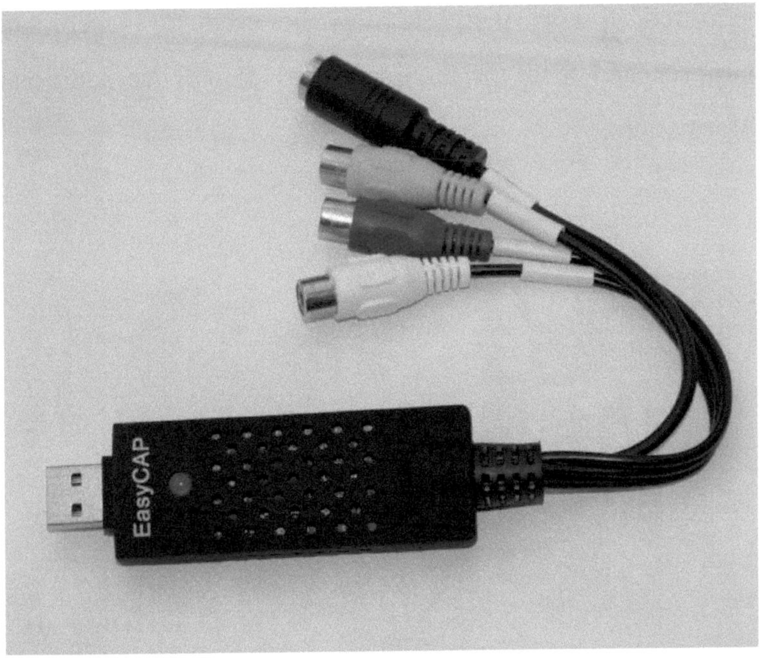

Figure 6: Frame Grabber with USB- interface.

With the GPS and also the altitude information via the air pressure sensor, it would also be possible from a technical point of view to control the multicopter solely via the screen. However, this is not permitted for legal reasons, because there must always be visual contact with the octocopter. This line of sight rule is mandatory in many countries. There must therefore always be visual contact with the remote-controlled aircraft. Thus, control via the ground station alone is not permitted. It may therefore only be used for supervision purposes and also for recording and displaying additional information.

Figure 7: FPV- camera, mounted on the octocopter.

In any case, it is also possible to involve a second person. This can read the data from the ground station and provide an assistance to the multicopter pilot. This can be information about the battery voltage or flight time, for example, or the evaluation and comparison of the position of the GPS with the background of the map information, or simply an explanation of which image and image quality the camera is currently delivering.

This type of very technology-intensive model building goes far beyond remote-controlled model aircraft building. Before starting, some tests are necessary until all transmissions of the entire system work. So a total of three transmissions are running in parallel, these are the 2.4 GHz of the remote control, then the 5.8 GHz of the video transmission and also the telemetry, which in the example works at 433 MHz. This type of model building may not be in the interest of everyone. But it is also the good property of today's technology that it offers all these possibilities. You can use them, but you don't have to.

4. Video glasses

If you really want to experience your model completely isolated from the rest of the environment, video glasses are the first choice. With them you can only experience the transmitted images or let them work on you. Here, too, the legal aspects must be observed in practice, because such a flight is not allowed outdoors because it is not a 'line of sight'.

However, it is always possible that the quadcopter pilot controls his aircraft with the remote control and someone else puts on the video glasses.

Video glasses have one or two color LC displays installed. A typical image resolution is the HDMI standard. Receivers with antennas are often built directly into the glasses so that the images can be transmitted directly from the model to the glasses. There are also variants in which the receiver is connected externally.

In the simplest and also most common variant, the images are not transmitted in stereo, but in mono. This also has the advantage that only one camera has to be installed on the quadcopter.

Figure 8 shows video glasses with an integrated receiver. In very sophisticated systems, the video glasses even contain an IMU, an 'Inertial Measurement Unit'. This can be used to measure whether you turn or tilt your head. The cameras are then mounted with gimbals on quadrocopters or multicopters. When this information is coupled, then they can move in the same direction as one turns or tilts one's head. This gives you the impression that you are sitting in the model are steering it by yourself.

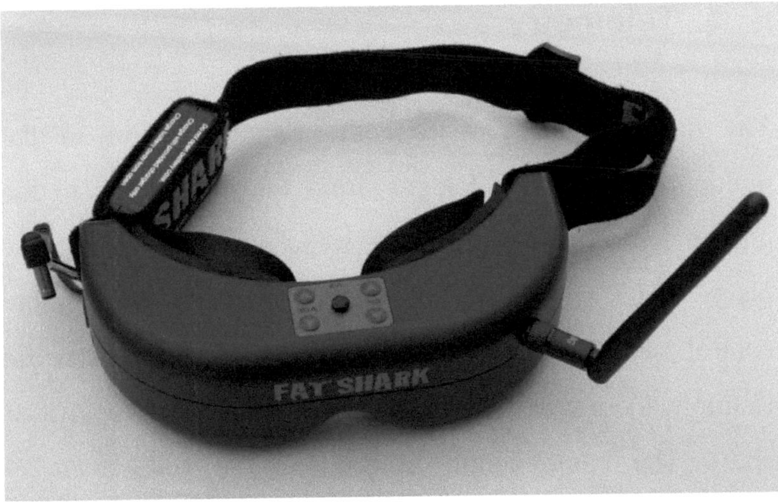

Figure 8: Video glasses with integrated receiver.

The above considerations show that both the cameras and video glasses can have very different properties. The cameras can be mounted rigidly or movably with a gimbal. The video glasses can be designed as mono or stereo images and some can even provide feedback on head movement.

As a beginner, you should first buy a simple mono video glasses. You can get good first experiences with it. If this makes you even more interested in FPV, you can also test the other extensions. This goes all the way to 3D imaging with a motorized gimbal for rotating and swiveling cameras.

5. Telemetry

The data transmission from the remote control to the multicopter and the telemetry from the multicopter to the ground station take place at the same time and in the same place. That's why it makes sense if telemetry and remote control send on different frequency bands. In the specific example, these were carried out in the same way as described above. The remote control transmits on the well-known 2.4 GHz and the telemetry on 433 MHz. Figure 9 shows the telemetry transmitter. It is plugged directly into the flight controller.

Figure 9: Telemetry transmitter with 433 MHz.

6. Ground station

In the present example, the telemetry data is read out directly from a receiver that can be plugged into the PC via the USB interface. The transmitter and the receiver are in a permanent data exchange and mutual communication. Therefore, strictly speaking, they are two identical devices. The telemetry transmitter shown above in figure 9 also serves as a receiver in the identical version at the ground station.

The data is now displayed using the PC software 'Mission Planner' in the practical example of the octocopter. In addition, the flight controller itself is also equipped with a memory card. This also allows so-called log data files to be stored locally on the multicopter with the chronological progression of a wide variety of parameters. In the event of an error, the data can then provide valuable information and tips about the positions, speeds and accelerations in all directions, the power consumption, the battery voltage, the PWM signals for the brushless controllers, GPS signals and much more. But that only works offline, i.e. after the flight. An online data storage at the ground station is also possible, using the transmitted telemetry data.

The telemetry data, which is transmitted online, i.e. during the flight, can be further processed on the PC in various ways. In this case, the 'Mission Planner', together with the PC, serves as a ground station. It can also display some selected parameters directly on the screen. This can be roughly seen in figure 10. For example, the current location of the multicopter with the transmitted GPS signal is shown directly together with the map background. The battery voltage or the current altitude in relation to the take-off point are also important information about the condition of the aircraft. And the image from the onboard camera is also displayed via a separate video receiver or one that is built into the video glasses.

Figure 10: Telemetry data and pictures at the ground station.

Figure 11: Visualization of a flight with waypoints.

Figure 11 shows the evaluated data of a log file. The image shows the recorded GPS data of a model airplane.

The previously programmed waypoints are then flown using an internal control algorithm with the flight controller. It tries to reach the waypoints as close as possible with a GPS control. Tolerances are also possible so that the routes can be flown optimally. These are shown in figure 12 as circles around the waypoints.

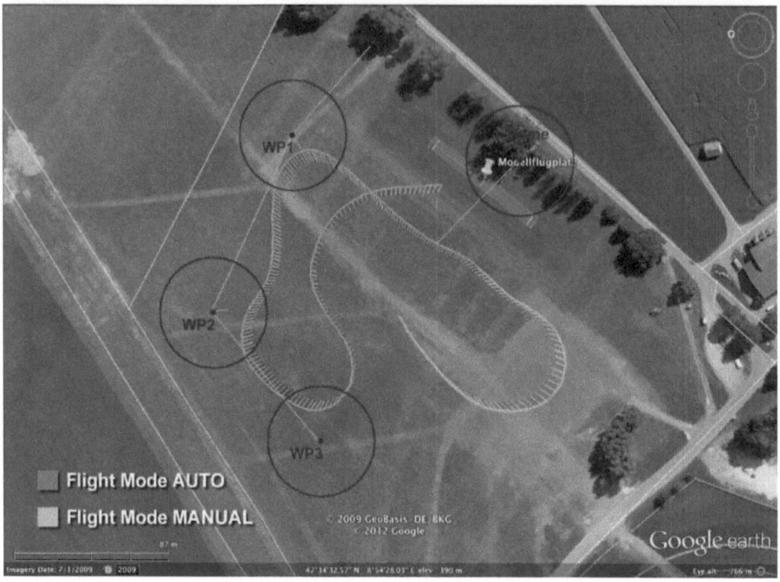

Figure 12: Waypoints and tolerance circles.

For the 'Auto' flight mode, so-called waypoints were programmed into the flight controller. Together with the GPS system, the model aircraft can fly off these waypoints independently (flight mode 'Auto'). Due to the regulation of visual flight, it is possible with the flight controller and also required by law that you can also switch to the 'Manual' flight mode at any time.

7. Literature

[1] Schenk, Helmut. "Der Standschub von Propellern und Rotoren." (2007).

[2] Wendel, Jan. *Integrierte Navigationssysteme: Sensordatenfusion, GPS und Inertiale Navigation*. Oldenbourg Verlag, 2009.

[3] Grewal, Mohinder S., Lawrence R. Weill, and Angus P. Andrews. *Global positioning systems, inertial navigation, and integration*. John Wiley & Sons, 2007.

[4] Büchi, Roland. *Radio control with 2.4 GHz*. BoD–Books on Demand, 2014., ISBN 978-3732293407

[5] Passern, Ulrich. *Das LiPo-Buch: Grundlagen und Praxistipps-Aktualisierte Neuauflage*. Verlag für Technik und Handwerk, 2016.

[6] Föllinger, Otto, Frank Dörrscheidt, and Manfred Klittich. "Regelungstechnik. Einführung in die Methoden und ihre Anwendung." *Berlin: Elitera Verlag* (1978).

[7] Büchi, Roland. "Brushless-Motoren und–Regler. 1." *Auflage, Baden-Baden: Verlag für Technik und Handwerk neue Medien GmbH* (2013).

[8] da Silva LR, Flesch RC, Normey-Rico JE. Controlling industrial dead-time systems: When to use a PID or an advanced controller. ISA transactions. 2020 Apr 1;99:339-50.

[9] Unbehauen H. Regelungstechnik. Braunschweig: Vieweg; 1992.

[10] Buchi, Roland, et al. "A remote controlled mobile mini robot." *MHS'96 Proceedings of the Seventh International Symposium on Micro Machine and Human Science*. IEEE, 1996.

[11] G. Schwarze: Bestimmung der regelungstechnischen Kennwerte von P-Gliedern aus der Übergangsfunktion ohne Wendetangentenkonstruktion, In: - messen-steuern-regeln Heft 5, S. 447-449, 1962